APHIS	Animal and Plant Health Inspection Service
BDRD	Biological Defense Research and Development
BIWAC	Biosurveillance Indications and Warning Analytic Community
BSP	Biosurveillance Portal
BST WG	Biosurveillance Science and Technology Working Group
BSV	Biosurveillance
CAPS	Cooperative Agricultural Pest Survey
CDC	Centers for Disease Control and Prevention
DHS	Department of Homeland Security
DoD	Department of Defense
DOI	Department of the Interior
DTRA	Defense Threat Reduction Agency
EIP	Emerging Infection Programs
EROS	Earth Resources Observation and Science
ESSENCE	Electronic Surveillance System for Early Notification of Community-based Epidemics
FDA	Food and Drug Administration
HHS	U.S. Department of Health and Human Services
HSPD	Homeland Security Presidential Directive
ICLN	Integrated Consortium of Laboratory Networks
LRN	Laboratory Response Network
NAHLN	National Animal Health Laboratory Network
NAHMS	National Animal Health Monitoring System
NBIC	National Biosurveillance Integration Center
NBIS	National Biosurveillance Integration System
NCMI	National Center for Medical Intelligence
NEON	National Ecological Observatory Network
NGDS	Next Generation Diagnostics System
NIH	National Institutes of Health
NOAA	National Oceanographic and Atmospheric Administration
NPDN	National Plant Diagnostic Network
NSF	National Science Foundation
NSIV	National Swine Influenza Virus Surveillance Program
NSTC	National Science and Technology Council
NWHC	National Wildlife Health Center
OSTP	Office of Science and Technology Policy
PHAA	Public Health Actionable Assays
PHEMCE	Public Health Emergency Medical Countermeasure Enterprise
POC	Point of Care
POCTRN	Point of Care Technologies Research Network
PON	Point of Need
R&D	Research and Development
RAPIDD	Research and Policy for Infectious Disease Dynamics
Roadmap	Biosurveillance Science and Technology Roadmap
S&T	Science and Technology

Strategy	National Strategy for Biosurveillance
USDA	U.S. Department of Agriculture
USGS	U.S. Geological Survey

Table of Contents

The *National Strategy for Biosurveillance (Strategy)*, published in July 2012, calls for "a coordinated approach that brings together Federal, state, local, and tribal governments; the private sector; nongovernmental organizations; and international partners" to enhance existing biosurveillance capabilities and, where necessary, develop new ones that provide decision makers and responders with the essential information they need to mitigate impacts of threats to health and associated economic, societal, and political consequences. The *Strategy* recognizes that a well-integrated national biosurveillance enterprise can saves lives by providing essential information for better decision making at all levels.

The interagency Biosurveillance Science and Technology Working Group (BST WG), chartered in May 2012, was established under the Subcommittee on Biological Defense Research and Development of the National Science and Technology Council's Committee on Homeland and National Security to develop national biosurveillance research and development priorities to enable the Core Functions of the *Strategy*.

The BST WG established four sub-working groups, each focused on addressing a distinct but interrelated aspect of biosurveillance science and technology: aberration detection; risk anticipation; threat identification and characterization; and information sharing, integration, and analysis. Together, the four groups identified the following high-priority research and development objectives to be prioritized in this period of limited resources:

- Establish baseline levels of community and ecosystem risks, threats, and health;
- Identify causes of aberrations from normal at the ecosystem, organism, reservoir, vector, and host nexus;
- Identify indicators that are associated with potential outbreaks and develop models using these indicators to assist in better decision making at all levels;
- Enhance information integration, analysis, and sharing platforms for improved situational awareness of biosurveillance information at all levels, including with international partners, as appropriate;
- Further develop technological solutions that integrate and analyze electronic health information, while protecting private information, to better inform health decision making;
- Identify and evaluate the utility of novel sources of biosurveillance information, such as social media;
- Improve exposure assessment and diagnostic capability, especially at the point of care, to enable accurate and timely collection of information for early detection and situational awareness throughout an incident; and
- Improve identification and characterization of known and unknown health threats.

Achievement of these high-priority research and development objectives and realization of the integrated biosurveillance enterprise called for in the *Strategy* will require coordination across Federal organizations, academia, industry, and the international community.

Chapter 1

Introduction

"A well-integrated, national biosurveillance enterprise is a national security imperative."

–National Strategy for Biosurveillance, July 2012

Background

Threats to human, animal, and plant health have the potential for significant economic, social, and political consequences. These consequences can be minimized by quick, well-informed action; the sooner a threat is detected and understood, the faster a response can be mounted and the threat's effects minimized. The *National Strategy for Biosurveillance (Strategy)*, published in July 2012, calls for "a coordinated approach that brings together Federal, state, local, and tribal governments; the private sector; nongovernmental organizations; and international partners" to enhance existing biosurveillance[1] capabilities and, where necessary, develop new ones that provide decision makers and responders with the essential information they need to mitigate impacts of threats to health—including the health of people, plants, domestic animals, and wildlife—and to food security and agriculture.

The *Implementation Plan for the National Strategy for Biosurveillance (Implementation Plan)* identifies and prioritizes actions needed to achieve the goals of the *Strategy*. One such action is the development of "…a research and development (R&D) roadmap that encourages innovation and collaboration in priority R&D areas…and addresses key scientific and technological gaps to strengthen the biosurveillance enterprise."

This National Biosurveillance (BSV) Science and Technology (S&T) Roadmap (*Roadmap*) identifies and prioritizes the R&D efforts needed to provide decision makers at all levels with the accurate and timely information needed to develop effective responses to incidents that threaten health. The R&D objectives in this *Roadmap* are designed to facilitate the accomplishment of the core functions and actions identified in the *Strategy* and *Implementation Plan*, respectively. Consistent with the *Strategy* and Homeland Security Presidential Directive (HSPD)-21 entitled *Public Health and Medical Preparedness*, this *Roadmap* focuses on S&T for anticipating significant health incidents involving naturally occurring, accidental, or manmade threats; rapidly and accurately identifying and characterizing incidents that occur; and effectively integrating, sharing, and analyzing the information available at each stage. Achieving the S&T objectives in this *Roadmap* will permit better decision making during an incident, resulting in improved mitigation, response, and recovery[2] that may ultimately save lives and improve health.

This *Roadmap* also identifies areas of potential department and agency collaboration to better leverage existing funding streams.

---

[1] Biosurveillance is the process of gathering, integrating, interpreting, and communicating essential information related to all-hazards threats or disease activity affecting human, animal, or plant health to achieve early detection and warning, contribute to overall situational awareness of the health aspects of an incident, and to enable better decision making at all levels.

[2] S&T to enable improved mitigation, response, and recovery is described in a separate S&T Roadmap in production by the Biological Response and Recovery Working Group. Development, efficacy, and recommended use of post-exposure prophylaxis and/or other medical countermeasures are addressed in the 2012 HHS *Public Health Emergency Medical Countermeasures Enterprise Strategy* and *Implementation Plan*.

Methodology

To create this *Roadmap*, the BST WG was established under the Subcommittee on Biological Defense Research and Development of the National Science and Technology Council's Committee on Homeland and National Security. The BST WG established four sub-working groups to explore four S&T focus areas in support of the *Strategy's* Core Functions:

Aberration detection – Define and prioritize R&D needed to establish the baseline condition of the environment and/or human (including vulnerable subpopulations), animal, or plant populations that is sufficiently robust to permit rapid identification of aberrant incidents to drive preparedness and timely, focused investigation.

Risk anticipation – Define and prioritize R&D needed to identify antecedent conditions and characterize complex interactions that permit prediction of an impending natural or intentional incident and to forecast impacts from such incidents.

Threat identification and characterization – Define and prioritize R&D needed to ensure exposures and health threats are identified rapidly and accurately and can be sufficiently characterized to provide needed information to decision makers, including responders and healthcare providers.

Information integration, analysis, and sharing – Define and prioritize R&D needed to enable improved integration, sharing, and analysis of BSV data in near real-time and in a format that provides essential information to decision makers, including responders and healthcare providers.

Each of the four sub-working groups was tasked to develop a baseline of the current state of BSV programs relevant to their focus area. This baseline served as a foundation to conduct a gap analysis with respect to the *Strategy* and identify priority S&T needs. Rather than describe all ongoing BSV-related programs identified by this process, this *Roadmap* highlights some representative Federal Government activities within each focus area.

Although many of the concepts and gaps described in this *Roadmap* have application across all hazards, this report focuses on biological threats to national security, including known and emerging infectious disease agents with potential to significantly affect the health of the environment, plants, animals, and humans around the globe, whether naturally occurring or released accidentally or intentionally.

Chapter 2
Foundations: Aberration Detection

Key Research Priorities:

- ❖ *Establish baseline levels of community and ecosystem risks, threats, and health*
- ❖ *Enhance methods and tools to rapidly detect aberrations from the baseline*

Background

Watching for and detecting aberrations from baseline conditions are the first essential steps to assessing and anticipating health risks and threats; identifying and characterizing those threats; and integrating, analyzing, and sharing information about the threats, as called for in the *Strategy*. Effective decision making relies on detecting aberrations from baseline conditions (e.g., a rapid increase in the number of influenza cases) sufficient to warrant a threat alert and the recommendation of precautionary or preventive measures (e.g., attention to hand washing and vaccination for influenza or mass prophylaxis for a bioterrorism event).

Aberration detection requires an ability to reliably distinguish a valid signal from background noise and qualitatively characterize and interpret the significance of that signal. Of course, in order to detect potentially threatening aberrations from "normal conditions" it is necessary to know what is "normal"—the baseline state of the environment or of human, animal, or plant populations against which aberrations can be measured. Moreover, this capacity must be dynamic and context-sensitive; what is normal now or in one circumstance may not be in the future or under different circumstances.

Current approaches to surveillance have variable accuracy and timeliness and in some cases detect aberrations only quantitatively without the necessary qualitative characterization. Most quantitative detections require expert evaluations before action should be taken. For example, syndromic surveillance systems may identify illness clusters early (before diagnoses are confirmed) and alert public health agencies, but the systems do not necessarily determine the cause, thus requiring additional subject-matter expertise to make sense of a given alert and drive subsequent action. Qualitative characterization is straightforward if the surveillance data are specific (e.g., results from influenza A laboratory tests), but can be quite limited if the data are non-specific (e.g., spike in emergency department visits for fever or symptoms that are shared by multiple diseases). Given the current prevalence of non-specific surveillance data, the development of methods for qualitative characterization is critical. Thus, any efforts to improve aberration detection should go beyond statistical improvements in detection accuracy to include qualitative analysis of aberrations.

Current Programs

Multiple Federal departments and agencies are working toward improved aberration detection, with some programs focused in particular on characterization of normal background. These efforts include the following:

Human Microbiome Project: The National Institutes of Health's (NIH) Human Microbiome Project[3] aims to characterize microbial communities found at multiple human body sites (e.g., nasal passages, oral cavities, skin,

---

[3] http://commonfund.nih.gov/hmp/

gastrointestinal tract, and urogenital tract) and look for correlations between changes in the microbiome and human health and disease. Initiatives include developing reference sets of microbial genome sequences; characterizing through molecular approaches the relationship between changes in the human microbiome and human health and disease; development of new technologies and tools for computational analysis; and establishment of resource repositories (e.g., reference strains, data, software).

National Ecological Observatory Network (NEON): The National Science Foundation (NSF) is supporting the construction and operation of NEON is a continental-scale ecological observation and research platform designed to enable the assessment of both the human causes and biological consequences of environmental change. NEON will provide data and data products needed to improve our understanding of and ability to forecast the impacts of climate change, land use change, and invasive species on ecosystem structure and function – specifically including biodiversity, biogeochemistry, ecohydrology, and infectious disease. These data and high-level data products will be available in close to real-time and will provide a regional to continental scale environmental baseline and the ability to detect changes from that baseline.

National Wildlife Health Center (NWHC): The Department of the Interior (DOI) U.S. Geological Survey's (USGS) NWHC conducts long-term monitoring and surveillance for wildlife health and disease detection and prevention in both aquatic and terrestrial environments. The NWHC provides diagnostic and disease research on wildlife; leads diseases investigations; participates in emergency response to medium and large-scale wildlife die-offs; and provides education, training, and outreach to wildlife personnel at both national and international levels.

National Animal Health Monitoring System (NAHMS): The U.S. Department of Agriculture's (USDA) Animal and Plant Health Inspection Service (APHIS) NAHMS[4] has been in place for over 20 years. The NAHMS Program Unit conducts national studies to provide essential information on livestock and poultry health and management to decision makers, including producers, researchers, and policymakers. Each animal group is studied at regular intervals, providing up-to-date and trend information needed to monitor animal health, support trade decisions, assess research and product development needs, answer questions for consumers, and set policy.

Capability Needs

The following capabilities are needed to strengthen aberration detection:

- **Surveillance methodologies that integrate traditional monitoring (i.e., pathogen, environmental, health) with background data (i.e., meteorological and population dynamics) that may influence risk**

  New modeling and ecological forecasting approaches have the potential to enhance the effectiveness of current strategies for predicting the likelihood of disease outbreaks and determining likely impacts when a threat is detected. For example, Hantavirus is endemic in the southwestern United States, where it is carried by deer mice. Research into the factors that determine the population dynamics of mice, such as multiyear patterns of rainfall, now enable public health agencies to issue warnings in years where the risk of human infection is high because of predicted increases in mouse populations. While imperfect, this integrated surveillance has the potential to dramatically reduce human infections as well as improve the speed of diagnosis and efficacy of treatment.

- **Ability to detect early warning signs of changes occurring at different and changing spatial and temporal scales**

  Automated aberration detection approaches often depend on data extraction and/or translation of foreign language reports—endeavors that would be greatly improved by

---

[4] http://www.aphis.usda.gov/animal_health/nahms/

advances in computer science and, in particular, the field of statistical anomaly detection in low signal-to-noise environments. While improvements are made in these areas of basic science, BSV could benefit from lessons learned by researchers in other fields that also rely on weak-signal extraction, such as sonar and radar detection of submarines and missiles.

- **Advancements in data sharing and integration and communication technologies including assessment methods**

  Equally important to the detection of aberrations is the need to integrate and analyze the monitored data and communicate the information to decision makers rapidly and securely. For example, early recognition of disease through the examination of data from electronic medical records has public health value only if that signal is available to decision makers in time to influence a response that can mitigate the impacts of the disease. Cooperation among Federal and non-federal stakeholders, including the scientific community and public and private healthcare providers, is essential to achieve an efficient and reliable surveillance system.

- **Education and training to meet anticipated needs for BSV professionals**

  New and non-traditional data collection and processing techniques have transformed the art and science of epidemiologic analysis. That in turn has increased the need for a more specialized workforce skilled in the application of these new and emerging capabilities. In order to prepare the next generation of scientists and engineers for successful careers as BSV professionals, education and training opportunities should be developed.

Research Priorities

Based on the capability needs described above and analysis of current programs, the following broad research priorities are proposed, with accompanying specific objectives:

- **Establish baseline levels of community and ecosystem risks, threats, and health**
  - Assess baseline exposures to endemic, occupational, and environmental threats;
  - Expand spatial and temporal mapping of endemic and epidemic disease;
  - Determine levels of immune (natural and vaccine-derived) protection for various populations;
  - Understand human behaviors and animal behaviors associated with health and illness and demographic trends; and
  - Understand pathway issues associated with trade and movement of people (e.g., smuggling and unintentional transport of pathogens or pests with plant or animal products).

- **Enhance methods and tools to rapidly detect aberrations from the baseline**
  - Develop statistical or mathematical algorithms to quickly and reliably distinguish a valid threat signal from background conditions;
  - Improve language processing and parsing tools;
  - Differentiate (temporal scale) between changes of immediate concern (e.g., influenza, foot-and-mouth disease) and those that have a large scale effect but spread more slowly (e.g., sudden oak death, fungal infections in bats and bees);

- Assess state-of-the-art in aberration detection methods and tools applied in other sectors (e.g., intelligence, financial markets, credit industry) and evaluate how they might be leveraged or applied to BSV applications; and
- Improve qualitative characterization of aberrations through targeted education and training of the next generation of BSV professionals.

Summary and Conclusions

Continued improvements in high-dimensional analysis of health data, integrated with data describing environmental factors, vector distribution, and other factors, can facilitate rapid detection and characterization. These approaches have considerable potential for predicting, detecting, and characterizing aberrations in a way that could transform how surveillance data and related information is presented and used by decision makers. This reality underscores the need for an approach that rationally guides efforts to define a complex and dynamic baseline and focuses S&T investments on those anomaly detection methods and tools that will be most effective in achieving the goals in the *Strategy*. Similarly, improvements in the integration, analysis, and sharing of information among appropriately trained stakeholders can ensure that monitored data has the greatest impact on disease mitigation.

Chapter 3

Risk Anticipation

Key Research Priorities:

❖ *Sustain R&D aimed at improving understanding of determinants of disease emergence and reemergence*

❖ *Focus on R&D relating to forecasting technologies and models that consider ecological and evolutionary drivers of disease behavior*

❖ *Connect non-invasive data-gathering tools to other types of surveillance data to improve the ability to detect antecedent conditions and the earliest indications of a significant incident*

Background

Globalization and ecological pressures are increasing the risk that a biological incident of national significance, such as the emergence of a novel infectious agent and/or a global pandemic, will occur. An important component of mitigating the consequences of such an incident is to anticipate when and where it may occur, enabling a more timely and well-informed response. The earlier a risk is anticipated and an actionable warning is broadcast, the greater the likelihood that a health threat can be prevented or its impacts significantly reduced. Risk anticipation requires the means to predict when an incident is likely to occur and to forecast its potential impacts on a given population. Technologies that help anticipate the risk seek to identify, measure, and analyze the many factors that set the conditions for, or directly influence, emergence of threats.

Emergence of many natural diseases is directly affected by easily measured environmental conditions, such as when isolated host environments are encroached upon. However, the complex patterns and interactions among hosts, vectors, the pathogens themselves, and the environment are not easily elucidated. For example, pathogenic determinants in the microbial world develop as a result of myriad biological and ecological pressures. The interaction between microbes and human and animal populations is affected by climate, the presence of pharmaceuticals, chemical additives and compounds, and social behaviors. Changes caused by those interactions occur with varying speed and scope to set the conditions for disease emergence. In addition, there remain a number of uncertainties about which pathogenic changes act to increase virulence, host range, or other factors. While there are ongoing R&D projects to understand these complex interactions, realizing the *Strategy's* goals will require strategic integration of efforts including near- and long-term S&T investments.

Current Programs

As described in Chapter 2, multiple Federal departments and agencies fund efforts to enhance baseline knowledge of genetics, molecular biology, proteomics, ecology, and epidemiology of known and emerging threats affecting the ecosystem, humans, animals, and plants to determine their relevance to national security and health. Traditionally, research has focused on retrospective evaluation to better understand data requirements and identify determinant variables from existing outbreaks. In order to prospectively assess current sparse data and provide projections of potential disease risks in an anticipatory manner, both pre- and post-incident, research priorities must be better articulated and coordinated among government, academia, industry, and non-governmental organizations.

Infectious Disease Emergence and Transmission: The joint NSF/NIH Ecology and Evolution of Infectious Disease program[5] supports efforts to understand the underlying ecological and biological mechanisms that govern relationships between human-induced environmental changes and the emergence and transmission of infectious disease. Similarly, NIH and the Department of Homeland Security (DHS) support a cooperative program called Research and Policy for Infectious Disease Dynamics (RAPIDD)[6]. Through an extensive series of workshops, working groups, and postdoctoral fellowships designed to address critical challenges, RAPIDD seeks to understand which models and modeling approaches will facilitate adequate operational capacity; how models relate with one another and with data of various quality and scale; and how the needs of decision makers can be characterized and addressed through modeling. RAPIDD focuses on improving the modeling of foreign animal diseases and zoonotic infections, aiming to make models more reliable and relevant to policy makers preparing for or responding to outbreaks. Finally, the NIH Models of Infectious Disease Agent Study is a transdisciplinary consortium of research groups with the mission of developing computational, mathematical, and statistical models of infectious disease dynamics and assisting decision makers to prepare for, detect, and respond to infectious disease threats.

Current technologies and tools to recognize and visualize early indicators include remotely-sensed imagery, common operating pictures, and disease modeling/forecasting. Federally funded modeling research on specific, highly infectious diseases, such as novel respiratory diseases with pandemic potential, is currently most useful for mitigation and response. Some success has been achieved in computationally assisted methods for predicting disease outbreaks. This success remains in the realm of disease risk mapping (e.g., ecological niche modeling), typically of vector-borne disease where meteorological and environmental conditions are the principal drivers and the diseases recur with some regularity in the same general location (e.g., cholera, Rift Valley fever, Hantavirus). In contrast, the current state of science does not allow for accurate predictions of the emergence of novel diseases or of the reemergence of diseases for which no regular cycle of emergence is known. Satellite instruments are powerful non-invasive data-gathering tools and are currently the most suitable tools to identify environmental and climatic precursors to infectious disease outbreaks. Current satellite operations may make it possible to identify early (or predictive) indications of biological threats by capturing changes in the relationships among human, animal, and plant populations, such as population density, interactions and migration, and environmental conditions. Although these efforts present promise in the near-term for informing decision making, true predictive capability will require maturation of the science behind current modeling methods and continued collaboration among government, academia, industry, and non-governmental organizations.

Global Weather and Climate Models: The National Oceanographic and Atmospheric Administration (NOAA) National Centers for Environmental Prediction runs global weather and climate models and makes global weather and climate forecasts. It leverages its expertise in the use of model and monitoring data to guide informed decision-making about human diseases by, for example, investigating the relationship between climate and incidence of meningitis. NOAA's Center for Satellite Applications and Research—part of the National Environmental Satellite, Data, and Information Service—develops health-related satellite data sets and products to help inform decisions relating to such diseases as malaria and dengue fever.

---

[5] http://www.fic.nih.gov/programs/Pages/ecology-infectious-diseases.aspx
[6] http://www.fic.nih.gov/about/staff/pages/epidemiology-population.aspx

Earth Resources Observation and Science (EROS) Center[7]: In partnership with the National Aeronautics and Space Administration, the USGS's EROS Center has been the steward and distributor of Landsat satellite data for over 40 years. The Landsat series of satellite missions has collected imagery of the Earth's surface since 1972, providing the most comprehensive record of the global landmass ever assembled. The EROS mission includes developing, implementing, and operating remote-sensing-based land-change monitoring, assessment, and prediction capabilities needed to address science objectives at all levels – within the USGS, across the Federal Government, and around the world. Landsat data are extremely valuable for a range of applications that contribute to science, environmental monitoring, and homeland security.

Federal departments and agencies are also aggressively assessing the role of technologies, such as social media, in characterizing disease emergence, reemergence, and nefarious exploitation of biological agents. To date, however, these efforts are limited to intelligence data triage, post-incident analysis of signal-to-noise relationships, and veracity of reporting with an eye toward situational awareness.

National Center for Medical Intelligence (NCMI): The NCMI, a component of the Department of Defense's (DoD) Defense Intelligence Agency, develops structured intelligence warning approaches to disease events of national significance (intentionally caused or naturally occurring), leveraging disparate open-source data, intelligence derived from National Technical Means, epidemiologic expertise, and intelligence tradecraft to move the intelligence warning capability temporally closer to the origins of an incident with the ultimate goal of warning prior to the emergence of a disease incident. Based on decades of basic data gathering and disease reporting, NCMI aims to expand its capacity to quickly identify abnormal disease occurrences and behavior globally. In partnership with the DoD Chemical Biological Defense Program and multiple Federal non-Title-50 partners, NCMI is exploring computational modeling, semantic fusion, entity resolution, link analysis, and molecular biological techniques that present some degree of promise in achieving a predictive capacity for the intelligence community.

PREDICT: PREDICT is a project of the U.S. Agency for International Development's Emerging Pandemic Threats Program that is building global surveillance to detect and prevent spillover of pathogens of pandemic potential that can move between wildlife and people. PREDICT has built a broad coalition of international partners to discover, detect, and monitor diseases at the wildlife-human interface using a risk-based approach that includes integrating digital sensing and on-the-ground surveillance at critical points for disease emergence. PREDICT is at the cutting-edge of recent technological advances allowing for rapid detection and diagnosis of high-risk viral families in all resource settings.

Capability Needs

Predicting the emergence of animal and plant diseases and zoonoses and forecasting the outcomes of their occurrence are significant scientific challenges. The principal needs to better forecast significant disease events and anticipate their risks include:

- **Understanding of antecedent conditions at appropriate spatial and temporal scales**

  The science of predicting discontinuous change in complex systems is relative immature—a challenge exacerbated by a relative paucity of data. In contrast to the physical sciences, in which systems and relevant variables can be enumerated and characterized in functionally closed systems, biological incidents occur in open systems, making the process of identifying and limiting relevant variables a daunting task.

---

[7] http://eros.usgs.gov/

- **Ability to forecast the dynamics of novel disease emergence or the exploitation of biological agents by adversaries**

  The ability to forecast impacts of the emergence of a novel disease or an intentionally released agent begins with having some fundamental understanding of environmental and behavioral factors that have predictive influence on an agent's behavior, including the movement of a human or animal population within the environment. Further, rapid characterization of the at-risk population, whether by direct knowledge (e.g., census data) or remote analysis (e.g., human geography/human terrain analysis), as well as the agent's transmissibility, mode of transmission, and generation time are key contributors to forecast impact. With respect to deliberate releases of pathogens, physical parameters of dispersal and the behavior of threat agents when aerosolized under various environmental conditions can illuminate the likely upper and lower bounds of the initial affected population. Basic epidemiology can then inform secondary transmission in the event an agent is contagious.

Research Priorities

Based on the capability needs described above and analysis of current programs, the following broad research priorities are proposed, with accompanying specific objectives:

- **Sustain R&D efforts aimed at improving understanding of determinants of disease emergence and reemergence, including ecological and evolutionary factors that promote the ability of organisms to move to new host-species (e.g., swine/avian influenza to humans) and acquire antimicrobial resistance**

  - Develop improved molecular approaches (genomics, meta-genomics, proteomics, non-destructive sequencing) to elucidate exposure causes and effects, and host/vector/reservoir-specific molecular-level relationships and interactions in the context of known and, ultimately, newly emergent biological threats;

  - Integrate the vast array of wildlife data (e.g., harmful algal blooms, fish kills, marine mammal strandings, migration) and environmental data (e.g., air quality, toxic exposures, public health surveys) collected by governmental agencies, academia, and industry, and merge these data into geographical and temporal visualization tools; and

  - Improve linkage between this basic research and the applied model-development research as described below.

- **Focus on R&D relating to forecasting technologies and models that consider ecological and evolutionary drivers of disease behavior**

  - Advance modeling R&D and analytical frameworks for specific classes of incidents (e.g., influenza versus cholera versus intentional release scenarios); and

  - Improve operational quality through rigorous testing of the reliability and validity of models.

- **Connect non-invasive data-gathering tools to other types of surveillance data to improve the ability to detect antecedent conditions and the earliest indications of a significant incident**

  - Characterize conditions, demographics, and environmental factors that can inform the earliest possible warning and projection of impact, as well as points of intervention to positively affect projected impacts;

- o Integrate emerging remote sensing capabilities/analysis (such as biological, chemical, and hyperspectral) with fixed, distributed autonomous or semi-autonomous surveillance platforms and conventional molecular biological tools to characterize and ultimately predict spatially and temporally important environmental variables that influence disease emergence within ecosystems, including humans;
- o Examine current coverage and capabilities of ground-based, *in situ* sensors for detecting threats, and enhance efficiency or expand, as appropriate and feasible;
- o Develop individual exposure assessment technologies, such as pre-symptomatic exposure biomarkers and individual dosimeters, to more rapidly and accurately detect exposures.

Summary and Conclusions

New foundational science and operational capabilities are needed to develop the ability to predict the impacts and risks posed by emerging diseases. Robust characterization of antecedent environmental conditions and the molecular dynamics of ecology-reservoir-host-pathogen interactions that can drive pathogenesis are essential to attain a true predictive capacity. Ultimately, such capabilities should be able to predict whether a health incident anywhere in the world will progress to one of regional, national, or international significance.

Chapter 4

Threat Identification and Characterization

Key Research Priorities:

- ❖ *Development of rapid, reliable next generation detection and diagnostic capabilities*

- ❖ *Development of new tools and methodologies to improve collection, preservation, transport, and preparation of clinical and non-diagnostic samples*

- ❖ *Development of instrumentation and large data-set-processing capabilities to rapidly identify characteristics of agents*

Background

Health is essential to national security. Health threats for human populations pose obvious security risks, but diseases in plant and animal populations can also pose serious security risks, including disruption of the food supply and commerce, with potentially long-lasting human health and economic impacts. Early detection of such threats can support determinations of the population exposed, diagnosis and treatment decisions, and containment of the threat. Strengthened laboratory and field capabilities are needed to recognize potential health threats as early as possible.

Diagnostic tests provide crucial information to surveillance programs in a variety of operational contexts, including U.S. laboratory networks and U.S. reference diagnostic laboratories associated with the World Health Organization, the United Nations Food and Agriculture Organization, and the World Organization for Animal Health (formerly known as the Office International des Epizooties). It is critical that technologies and methodologies to enable rapid and accurate identification and characterization of health threats are broadly available and fully functional.

Current Programs

A number of programs today contribute to the Nation's capacity to identify and characterize health threats. Several representative programs are described below, categorized by some of the overarching identification and characterization challenges they address.

- **Diagnosis of disease at point of care (POC) and point of need (PON):** Infectious disease diagnostic testing generally occurs in fixed-site clinical laboratories. Typically, clinical-laboratory-based diagnostic platforms are not cost effective or rugged enough for use in POC/PON settings where treatment can be initiated sooner. While several POC/PON tests exist for initial patient testing, they are limited by poor sensitivity or specificity and generally require a confirmatory clinical laboratory test result prior to treatment. Advanced POC/PON testing platforms with sensitivity and specificity similar to those used in clinical laboratories are becoming available for limited resource settings; however, they remain high complexity tests and continue to require a trained laboratory clinician to ensure reliable results, limiting their global adoption.

Point of Care Technologies Research Network (POCTRN): The NIH POCTRN drives the development of appropriate POC technologies through collaborative efforts that merge scientific, engineering, and technological capabilities with clinical need. The POCTRN is developing technologies for the diagnosis, screening, treatment, and monitoring of a variety of diseases. A major goal of POCTRN is higher quality care at reduced cost, with a shift in focus from specialized care for the treatment of late-stage disease to an emphasis on patient-centered approaches and coordinated care teams that promote wellness and effective disease management.

DoD Next Generation Diagnostics System (NGDS): The DoD NGDS is an incrementally acquired family of systems intended to provide an enhanced diagnostic capability from the laboratory to the field in support the DoD and its U.S. Government partners. Short-term investments will focus on the acquisition of a commercial off-the-shelf diagnostic system suitable for use in field and mobile laboratories and development of Food and Drug Administration (FDA)-cleared biological agent *in vitro* diagnostic assays. Long-term efforts will focus on the development of far-forward diagnostic capabilities and diagnostics for chemical, biological, radiological, and nuclear threats. Common themes throughout all increments of the NGDS program are alignment with therapeutics and the pursuit of common materiel solutions for endemic infectious disease diagnostics, and environmental sample analysis for public health and biological defense applications.

- **Detection of threat agents in non-diagnostic samples using field-based and laboratory systems:** A number of laboratory and disease-tracking networks are available to perform confirmatory testing of suspected disease threats (such as suspicious powders or animal droppings) and to detect disease trends in animals and plants. For example, the Centers for Disease Control and Prevention's (CDC) Laboratory Response Network (LRN) is an integrated system of state and local public health, Federal, military, and international laboratories that operate continuously for laboratory identification of threats associated with terrorism and other public health emergencies. Similarly, USDA and USGS reference laboratories are members of diagnostic laboratory networks such as the National Animal Health Laboratory Network[8] (NAHLN) and the National Plant Diagnostic Network (NPDN)[9], which provide rapid detection and confirmatory diagnoses of known threats to plants and animals.

Federally Funded Biological Detection Systems: There are several detection systems currently deployed to conduct continuous surveillance for biological threats in non-diagnostic samples by Federal departments and agencies and private-sector organizations. Examples include: DHS' BioWatch Program; DoD's Installation Protection Program and Joint Biological Point Detection System for wide-area detection of aerosol threats; the Environmental Protection Agency's Water Security Initiative to detect contamination of drinking water distribution systems; NOAA's Harmful Algal Bloom forecast network; the U.S. Postal Service Biohazard Detection System used to screen mail in U.S. postal facilities across the Nation; and DOI USGS' Migratory Wild Bird Disease Surveillance Program using migratory, genetic, and immunological data to identify likely routes of virus introduction, and prioritizing migratory bird species for sampling based on their potential to transmit highly pathogenic avian influenza into North America.

Public Health Actionable Assays™ (PHAA): The PHAA effort sets stringent assay performance evaluation and validation requirements for both non-diagnostic samples as well as clinical samples within the LRN member laboratories; PHAA ensures clinical samples are compliant with FDA regulations for *in vitro* diagnostic use. The evaluation and performance requirements were determined through an interagency effort. Assays that achieve this level of performance provide the high confidence identification results necessary in the public health community to inform early decision making.

Cooperative Agricultural Pest Survey (CAPS): USDA's APHIS leverages efforts by state departments of agriculture, universities, and industry partners in annual surveys targeting specific exotic plant pests that threaten U.S. agricultural and/or natural settings by funding a network of cooperators that participate in the CAPS program and conduct surveys and testing. The NPDN is a strong partner with Federal and state programs to provide a distributed nationwide network of public agricultural institutions to quickly detect high consequence pests and

---

[8] http://www.aphis.usda.gov/animal_health/nahln/
[9] http://www.npdn.org/

pathogens using federally validated methods and immediately report them to appropriate responders and decision makers.

National Swine Influenza Virus (NSIV) Surveillance Program: USDA and CDC initiated a swine influenza virus surveillance pilot project in 2008 to track the epidemiology and ecology of swine influenza virus in swine and human infections. Under this pilot project, when the human pandemic influenza outbreak occurred in April 2009, CDC quickly shared human H1N1 virus isolates with USDA, allowing USDA to rapidly develop, validate, and deploy to the NAHLN an A(H1N1)pdm09-specific diagnostic polymerase chain reaction assay. The surveillance pilot was expanded and then modified into the NSIV surveillance program to (1) monitor genetic evolution of swine influenza virus; (2) make swine influenza virus isolates available for research and an objective database for genetic analysis of these isolates and related information; and (3) select proper isolates for the development of relevant diagnostic reagents and vaccine seed stocks.

- **Rapid characterization of emerging and novel threats:** Laboratory characterization of a novel disease-causing agent involves the collection of high confidence genotypic and phenotypic information, such as those related to critical pathogen properties (e.g., phylogeny, virulence, growth, morphology, pathogenicity, viability, transmissibility, antibiotic susceptibility, and functional genomics); host immune response and early disease markers; microbial source identification indicators; and ecological adaptations. Data derived from this type of in-depth analysis informs medical countermeasure development and response.

Technologies for Rapid Characterization and Identification: CDC's Rapid Response & Advanced Technology Laboratory, in conjunction with the Biotechnology Core Facility and Office of Public Health Preparedness and Response, is facilitating an interagency multi-tiered technology approach for rapid pathogen detection, characterization, and identification. Applied R&D supports CDC/DoD collaborative efforts to operationalize Biomedical Advanced Research and Development Authority and Defense Threat Reduction Agency (DTRA) research investments and development of a systems approach for analysis of clinical samples for identification and characterization of known, emerging, and advanced biological threats. This includes genomic and proteomic technologies associated with multiplexed screening, microarrays, high-throughput sequencing, and bioinformatics analysis. This research also serves as proof of concept to address potential gaps in surveillance to detect natural or man-made changes in known agents and discover unknown/unexpected agents due to a reduced use of culture-based diagnostics.

- **Surge/follow-up capacity for clinical and non-diagnostic sample analysis:** Surge capacity is the operational capability to detect or diagnose threat agents in a large number of samples within a short time frame. The current U.S. surge capacity for emergency detection and diagnostic testing of samples generated after a potential incident relies on coordinated networking among cooperating laboratory systems. Coordination requires shared knowledge about specific laboratory sample preparation and analysis capabilities so that sample routing and results reporting can be optimized when surge capacity is required. Follow-up capacity relates to the fact that, in addition to technologies for predicting and detecting a disease outbreak, there is a need for surveillance methods specialized for use in the aftermath of an outbreak to determine the scope of the evolving threat and the level of continued risk to susceptible populations. Furthermore, novel surveillance efforts may be needed to certify that entities or areas are free of disease or to otherwise confirm exposure status for international trading partners or for other purposes.

Integrated Consortium of Laboratory Networks (ICLN): The ICLN is a DHS-chaired multiagency effort to bring together information, operations, and strategies from different laboratory systems for timely response to major incidents; participating networks include the CDC LRN, the NAHLN, the NPDN, the Food Emergency Response Network, the Environmental Response Laboratory Network, and the DoD Laboratory Network. The ICLN is a forum to share ideas, collaborate, and build relationships to support a more effective integrated response during emergencies (e.g., terrorist attacks, natural disasters, and disease outbreaks).

Capability Needs

The principal needs to improve the identification and characterization of threats include:

- **Improved sensitivity, specificity, and portability of multiplexed technologies capable of identifying, with confidence, known and unknown threats in complex samples**

  Developing capabilities for more rapid, accurate, and comprehensive early identification and characterization of emerging or re-emerging pathogens and toxic exposures, especially for those of unknown etiology, can reduce morbidity, mortality, transmission, and consequences. Moving diagnostic capabilities to POC/PON settings will result in faster initiation of treatment, as would developing a capability to identify patients requiring treatment earlier in the progression of symptoms. When possible, such technologies should be developed so they can be sustainably adopted by healthcare practitioners in low-resource international contexts and are amenable to adaptation to high-throughput situations. Focusing investments on POC/PON tests for which there is an available treatment, or where early identification of the disease can impact patient outcomes, will facilitate adoption and increase the return on investment.

- **Improved sample collection, preservation, transport, and preparation technologies and protocols**

  Current methods to collect, preserve, transport, and prepare clinical and non-diagnostic samples must be augmented to ensure presentation of high-quality samples for detection and diagnostic tools. This includes work to reduce the need for large sample sizes for testing, allowing for retention of as much of the original sample as possible for future evaluation.

- **Improved standards for testing and evaluation of detection tools, including the development of inclusivity and exclusivity threat-agent panels and validated testing and evaluation protocols**

  Confidence in detection tools is hampered by inconsistent expectations about the use and performance of these tools. This inconsistency is the result of varying standards for testing and evaluation methodologies and the use of reagents/materials that are not universally recognized as reference-quality. Improving test and evaluation standards will result in better information for decision makers and higher confidence in and acceptance of results.

- **Improved diagnostic technologies and access to signatures, reagents, strains, and sequence data, and an informatics and computational capabilities**

  To support development of diagnostic and detection platforms for known, emerging, re-emerging, and unknown pathogens, there is a need for a consortium of repositories with a central catalog of signatures, reagents (including libraries of both validated reference strains and nearest neighbor strains), and taxonomic and sequence data, as well as the informatics and computational capability to support warehousing, curatorial services, querying, and modeling of disparate types of relevant data. This repository and capability will provide the means for standardized technology development, validation of technology performance, and ability to interpret results.

- **Improved surveillance techniques, data sharing, and interoperability for plants, animals, and food**

  The ability to conduct efficient surveillance of health threats in plants, animals, and food is hampered by the lack of cost- and labor-effective methods to inspect a high percentage of these items; rapid field detection and diagnosis tools; and identification of pre-symptom or latent markers of infection or infestation. Furthermore, there are numerous databases of surveillance information in these sectors that are organized by species or industry, which contributes to a large technical gap in interoperability and minimal ability to share data from multiple disparate data streams, to inform when increased surveillance in a particular population is needed (e.g., when a spike in wildlife disease spurs increased surveillance of nearby livestock and poultry populations), or to establish a threat baseline to enable detection of emerging or re-emerging threats. It is particularly critical to improve foreign disease surveillance and data sharing capacity, and to enhance international capabilities to share critical disease surveillance data.

Research Priorities

Based on the capability needs described above and analysis of current programs, the following broad research priorities are proposed, with accompanying specific objectives:

- **Development of rapid, reliable detection and diagnostic capabilities**
  - Increase the speed and performance of threat detection, exposure, and disease diagnosis to support rapid and effective treatment decisions, contain disease, and mitigate the impact of a potential outbreak;
  - Move the determination of individual (asymptomatic) exposure and diagnosis of disease closer to the POC/PON setting, resulting in rapid initiation of treatment; and
  - Enhance global access to POC/PON tests.

- **Development of new tools and methodologies to improve collection, preservation, transport, and preparation of clinical and non-diagnostic samples, to include maintaining sample safety, integrity, and viability/culturability**
  - Ensure compatibility with nucleic acid, protein, and/or live organism characterization methods;
  - Integrate sample preparation technologies within detection or diagnostic tools, where possible, and reduce required ancillary equipment; and
  - Develop semi- to fully-automated sample collection, preservation, transportation, and preparation technologies that are compatible with current and emerging detection/diagnostic systems; are of sufficient scalability to support high throughput applications; and are deliberately developed to be sustainably adopted by healthcare practitioners in low-resource international contexts.

- **Development of instrumentation and large-data-set processing capabilities to rapidly identify characteristics of known agents, rapidly detect changes in known agents, and/or to discover the existence of unknown agents from samples in clinical or environmental matrices**

  o Examine characterization efforts (genotypic and phenotypic) for human, animal, and plant pathogens from high-risk areas around the world to detect emerging infectious diseases, and expand as appropriate and feasible;

  o Develop comprehensive databases linking pathogenic agents to disease outbreaks, location of origin, and other characteristics;

  o Establish a consortium of repositories of signatures, reagents, strains, and sequence data, and an informatics and computational capability for warehousing, curatorial services, querying, and modeling; and

  o Develop standardized processes to conduct "what if" analyses of the environmental and health impacts of a pathogen prior to patient identification and treatment initiation.

Summary and Conclusions

Currently, the array of accurate, durable, and reliable detection, measurement of exposures, diagnostic, and characterization systems and methods is inadequate. This insufficiency can lead to early confusion as to the nature, cause, and appropriate treatment of a health threat; reliance on expensive laboratory infrastructure for analysis; and a reactive rather than proactive response to incidents. Achieving the S&T goals in this area will increase the speed and accuracy of detection, exposure assessment, and identification, characterization, and disease diagnosis; increase confidence in identifying health threats; and provide situational awareness and critical information to support decision making associated with control and prevention, to include assessment of the potential economic and health consequences of an incident and rapid and effective treatment and control decisions.

Key Research Priorities:

- ❖ ***Development/enhancement of systems that improve near-real-time sharing of electronic health, diagnostic, and other anomalous health event data***

- ❖ ***Development of improved mechanisms to assess data/information sources for relevancy to BSV***

- ❖ ***Development of multilateral communication mechanisms among various levels of government, and the private sector (including healthcare providers, international partners, and others)***

- ❖ ***Development of a national, interagency BSV data-sharing framework that integrates data/information from disparate sources***

- ❖ ***Integration of all source data (intelligence, law enforcement, environmental, socio-economic, and health information)***

- ❖ ***Formalization of a means to effectively communicate uncertainty in BSV data used for decision making***

Background

The overarching goal of the U.S. BSV enterprise is the ability to make informed decisions earlier, enabled by analysis of near-real-time information and integration of numerous existing efforts at varying stages of development or deployment. Data sharing and integration, as described in HSPD-21, suggests "…international connectivity where appropriate, that is predicated on state, regional, and community level capabilities and creates a networked system to allow for two-way information flow between and among Federal, state, and local government public health authorities and clinical healthcare providers." A robust global system to coordinate the integration and analysis of information does not currently exist. Improved integration and analysis of information from multilateral capabilities would provide the Nation and international community with a powerful capability for early warning of an emerging incident and situational awareness while the incident is being characterized. The architecture for this framework would be inherently complex and its operationalization will require long-term R&D efforts and international collaborations.

Current Programs

Many novel and promising BSV programs have been developed for information sharing, integration, and analysis at the Federal, state, and local levels in recent years. In 2010, the Government Accountability Office summarized more than 100 Federal data sources and systems that could contribute to a national BSV enterprise.[10] Two efforts planned or currently underway to identify these systems include an expansion of CDC's previous internal efforts to capture non-CDC information systems as part of a proposed Federal BSV Registry and an effort funded by

---

[10] http://www.gao.gov/assets/310/306362.pdf

DTRA as part of its BSV program to identify, categorize, and assess the value of potential data streams relevant to BSV systems for DoD.

Extensive efforts have been made by numerous national and regional entities to use existing human electronic health records for syndromic surveillance, exemplified nationally by DoD's Electronic Surveillance System for Early Notification of Community-based Epidemics (ESSENCE) and the CDC BioSense system, and regionally by systems such as the Pennsylvania Real-time Outbreak and Disease Surveillance system and the National Collaborative for Bio-Preparedness. These systems leverage existing health data (e.g., administrative patient discharge coding data, laboratory data, and emergency department chief complaints) from participating sources for automated, near-real-time transfer and analysis of information for early detection of anomalous health events. These types of systems have demonstrated some success for situational awareness during an event, but their ability to provide reliable and timely early warning is yet to be established.

BioSense 2.0: BioSense 2.0 is a CDC chartered, community-guided public health surveillance system that provides the capability to expand the practice of syndromic surveillance at local, state, regional, and national levels. Hosted completely in a secure internet cloud computing environment, BioSense 2.0 is capable of rapidly monitoring outbreaks and harmful health effects of hazardous agents and tracking them throughout the duration of a public health emergency utilizing hospital emergency department record data. BioSense 2.0 provides state and local health jurisdictions storage, analysis, and aggregate data sharing capabilities through a syndromic surveillance platform. By 2014, BioSense 2.0 is projected to incorporate data from 65% of state and local jurisdictions, plus data from other healthcare sources. These data will contribute timely and accurate information to the overall situation awareness of local, state, regional, and national public health.

Standards for Biosurveillance Information Exchange: HHS previously sponsored the Healthcare Information Technology Standards Panel Biosurveillance Interoperability Specification that defined standards to promote BSV information exchange among healthcare providers and public health authorities. This effort helped establish a detailed data sharing framework for electronic health record information technology systems to encourage use of systems that would provide public health information directly to a public health authority.

A number of discipline-specific and surveillance partnerships exist. For example, the NIH's Influenza Research Database is an international and domestic collaborative research effort between the Centers of Excellence for Influenza Research and Surveillance and bioinformatics/genomics research programs with a focus on integrating diverse datasets and sharing data with the influenza virus research community. The Department of State implements foreign assistance projects in the Middle East and North Africa, South and Southeast Asia, Sub-Saharan Africa, and other regions, to promote safe, secure, and sustainable bioscience capacity that improves disease diagnosis, reporting, and response. International partnerships are critical to this work, and the United Kingdom, Canada, the Netherlands, and the Republic of Korea have each provided financial contributions to help advance U.S.-led bioengagement activities abroad.

National Biosurveillance Integration System (NBIS): NBIS is a national interagency BSV integration body coordinated by the DHS National Biosurveillance Integration Center (NBIC) in accordance with a series of U.S. laws and directives (HSPD-9 and -10, Public Law 110-53 Section 1101, Food Safety Modernization Act Section 205). NBIS member agencies integrate data within their BSV domain and share this information with NBIC after the data are analyzed by their subject matter experts. NBIC, in full collaboration with the NBIS, connects, correlates, and contextualizes information across domains through the production and dissemination of its analytic products. NBIC's integrating role enhances the Federal government's ability to provide early warning and shared situational awareness.

Emerging Infections Programs (EIP): The EIP are population-based centers of excellence established through a network of state health departments collaborating with academic institutions; local health departments; public health and clinical laboratories; infection control professionals; and healthcare providers. The EIP network's unique strength and contribution lies in its ability to quickly translate surveillance and research activities into informed policy and public health practice and to maintain sufficient flexibility for emergency response as new problems arise. Surveillance efforts of the EIP activities generate reliable estimates of the incidence of certain infections and provide the foundation for a variety of epidemiologic studies to monitor prevention strategies, explore risk factors, validate diagnostics and surveillance methods, and investigate spectrum of disease.

Biosurveillance Indications and Warning Analytic Community (BIWAC): The BIWAC is a self-organized, informal BSV information sharing group with participants from multiple U.S. government organizations. The BIWAC shares BSV data via unsophisticated web interfaces and has focused on interagency collaboration and relationship building.

The DoD Joint Science and Technology Office has initiated an R&D program, called the BSV "Ecosystem" program, which aims to heavily leverage private-sector innovations in data collection and analysis that have not been previously applied to BSV. It is envisioned that data from collection and analysis systems will eventually be linked in a cloud-computing construct to serve a range of analytical customers; however, a specific architecture has yet to be developed. The DoD Joint Program Executive Office for Chemical and Biological Defense is developing a BSV communications framework termed BSV Portal (BSP) that aims to provide a single web-based environment that will facilitate collaboration, communication, and information-sharing in support of the detection, management, and mitigation of man-made and naturally occurring biological events. While the current emphasis of the BSP is focused on a subset of DoD users, it is being developed with the capability for a broader user base in the DoD and interagency space and may serve as a component of a future national BSV enterprise.

Several Federal departments and agencies and offices have invested in BSV-related R&D, including HHS (CDC, NIH), DoD, DHS, NSF, the Department of Commerce (NOAA), and USDA. Most current BSV-related investments, however, support existing data integration systems to meet the day-to-day information needs of the organization. These existing efforts could form the basis for a national-level data integration enterprise with sufficient funds, mandates, and long-term plans.

Capability Needs

- **Ability to aggregate analyzed health data from different health sources, syndromic surveillance systems, and sectors to detect aberrations and discover spatial and temporal disease trends**

  While the potential for leveraging electronic information to augment disease surveillance is widely recognized, certain information-sharing criteria and research efforts need to be realized to advance this capability. Health data in electronic form have significant value for BSV since they provide an opportunity for more timely recognition of clinical signs in clusters of humans, animals, and plants that may provide an early indication of an emerging health incident. Effectively, appropriately, and securely sharing health event data, including parts of electronic patient records and laboratory data, has significant potential to improve national awareness of incidents that could progress to impact national security. However, a number of data-sourcing challenges exist. For example, data on human health can be collected from numerous types of patient management systems, laboratory information systems, and insurance systems, but there is a need for comprehensive terminology standardization across the spectrum of sources to ease data fusion and analysis while protecting the privacy of personal health data. Similarly, many livestock health records (domestic and foreign) are held by private industry and are not standardized from an informatics point of view, nor are they broadly accessible. Overall, there is a need for increased automated sharing and integration of electronic data from different health systems, different syndromic surveillance systems, and different health sectors.

- **Ability to determine what data/information sources are relevant to BSV**

  The BSV community needs a systematic mechanism to categorize, evaluate, and document the expected contributions and limitations of data sources useful for local, regional, national, or international BSV efforts including early warning, early detection, and situational awareness. A few typical examples of data sources include: clinical, syndromic, diagnostic, prescription, and over-the-counter drug use; news and social media; school and work absenteeism; Internet-based early warning clearinghouses; and aggregated commercial search engine query data. Each of those broad categories can include a range of specific data sources with vastly different performance metrics due to differences in quality, completeness, or timeliness attributes, or the methodology used to analyze them. This variation provides a rich pool of potentially useful BSV data; however, it also presents challenges, as it can be difficult to assess the relative value of each source. Further, these various types of data may be subject to differing privacy laws.

- **Sustained and appropriate multilateral information sharing**

  Producing a national BSV capability requires across-the-board collaboration of subject matter experts from different sectors (e.g., human, animal, plant, and environmental health; intelligence) at different levels of government, from the private sector, and with international partners. Active collaboration and information-sharing for BSV presents a number of challenges, especially since stakeholders have varying missions and roles, and the perceived requirements for early warning and situational awareness often necessitate novel or unconventional uses of preliminary information that cannot always be shared. Mechanisms to ensure rapid sharing (both "push" and "pull") of time-sensitive information aimed at supporting local, regional, national, or international decision- making can prove critical to achieving health and security goals. Confidentiality issues, the level of granularity to be shared among BSV systems, and actions mutually expected to result from shared information must be addressed and understood by participants in advance.

- **Integration of all existing and emerging BSV efforts and cross-domain information sources into a coherent BSV enterprise**

  One of the biggest challenges to successful national BSV is that, to be effective, information needs to be considered simultaneously from vastly disparate domains, including health, law enforcement, intelligence, environmental monitoring, remote sensing, international partners, and many others. Each domain has its own culture, investigative methodology, and tolerance for the timeliness and completeness of information necessary to make decisions and initiate response actions. Additionally, there are dozens and possibly hundreds of BSV initiatives and pilot projects that have been started at local, state, regional, and national levels. Without a universal system for integration of multiple, disparate data sources relevant to BSV, the synergistic value of bringing these data together will not be attained.

- **Standardized methods for determining the uncertainty in BSV data and clearly communicating the limits of the data and analytical techniques**

  For information to be actionable, some assessment and communication of its validity and accuracy is needed. BSV data present many analysis challenges because they can be inconsistently collected, reported, and analyzed; may contain collection and other types of data biases; can be derived from many disparate or incongruent sources; and may have gaps in temporal and spatial coverage. Each of these factors can add uncertainty to any analysis. In addition, local contextual knowledge is often needed to interpret the significance of any statistical deviation. These influences may or may not be known to the data owners and, as the data are analyzed by others, re-purposed, or otherwise reach a broader audience, the nuances related to specific data uncertainties may not be consistently and appropriately communicated. As a result, human expertise, which may be enhanced by technology, is key to a successful BSV enterprise. Research related to uncertainty quantification and communication is required to ensure decision makers are fully aware of the strengths and weakness of inferences that may be made as a result of BSV data analysis.

Research Priorities

Based on the capability needs described above and analysis of current programs, the following broad research priorities are proposed, with accompanying specific objectives:

- **Development/enhancement of systems that improve near real-time sharing of electronic health, diagnostic, and other anomalous health event data**

  o Develop systems for human, animal, and plant electronic health data with capacity for standardizing data elements to allow for potential interoperability of systems and data integration, including with international partners as appropriate;

  o Define a common "near-real-time" requirement for early incident detection, given the varying requirements among sectors; and

  o Develop standards to ensure secure, automated data transmission with rule-based sharing and role-based authentication.

- **Development of improved mechanisms to assess data/information sources for relevancy to BSV**

  o Continue efforts to define and establish mechanisms to assess data/information sources, particularly through coordination of DoD, HHS, and DHS efforts and with the rest of Federal interagency;

  o Develop and use metrics for each category of data source (e.g., sensitivity, specificity, timeliness) to assess the utility of tools, training programs, and strategies employed to support national and global BSV efforts; and

  o Establish a process and test data sets for calibration and system performance comparisons to assess whether a source is primarily beneficial for earliest possible detection, situational awareness, impact forecasting, and/or response efforts.

- **Development of multilateral communication mechanisms among various levels of government and the private sector (including healthcare providers, international partners, and others) to enable timely decision making at all levels**

- o Develop a community-endorsed and championed primary BSV Common Operating Picture, leveraging existing capabilities and lessons learned, to collect and share information at a national level and display a level of BSV information, including analyst insight, necessary to inform operational decision-makers across the government;
- o Evaluate ways to establish enduring, analyst-focused trust relationships across sectors and domains—and among international partners, organizations, and nongovernmental organizations—to facilitate a multi-institutional culture of information sharing; and
- o Establish mechanisms to securely transfer finished or unfinished information at different levels (e.g., open-source, sensitive, classified) among stakeholders and decision makers to enhance integration and usability.

- **Development of a national, interagency BSV data-sharing framework that integrates data/information from disparate sources to enable early warning and early detection of incidents and situational awareness during an incident**

  - o Integrate various existing and emerging BSV efforts and cross-domain information sources into a coherent BSV enterprise. This path forward may require a "system-of-systems" approach that capitalizes on successes and capabilities of the many existing systems by linking them in a new national BSV framework that does not currently exist; and
  - o Develop a viable approach for state and local authorities to access a national, interagency BSV data-sharing framework, as well as ways to collect their data.

- **Integration of all source data (intelligence, law enforcement, environmental, socio-economic, and health information) to enhance the detection of a disease event and facilitate warning and forecasting of impact**

  - Continue development of automation tools that allow efficient knowledge discovery through exploitation of existing large and massive data sets, particularly unstructured massive information stores such as PubMed, the National Center for Biotechnology Information, and other National Library of Medicine resources; as well as social media; online engine searches; and traditional media reporting; and
  - o Establish an interagency BSV information technology development, information management, and knowledge generation coordination initiative.

- **Formalization of a means to effectively communicate uncertainty in BSV data used for decision making**

  - o Develop standardized methods for determining the uncertainty in BSV data and clearly communicating the limits of the data and analytical techniques.

Summary and Conclusions

A major challenge to successful national BSV is that information must be considered simultaneously from vastly disparate domains, including health, law enforcement, intelligence, and international partners. Sharing of this information is limited, due to real or perceived confidentiality and other issues, and is often not possible. Integration of multilateral capabilities will enable strengths and resources to be leveraged, providing a powerful capability for early warning and situational awareness. Successful implementation of a national BSV enterprise will support a more comprehensive national information sharing capability to save lives and reduce illness.

Meeting the challenge of establishing the national BSV enterprise called for in the *Strategy* will be difficult, but is not insurmountable. Progress has already been made and current programs and technologies are making strides toward meeting the *Strategy's* goals. Continuing this progress will require focusing investments and coordinating efforts across the Federal Government on enabling S&T, with participation from academia, industry, and international partners.

Investing in the goals identified in this *Roadmap* will increase the speed and accuracy of disease detection, identification, characterization, and information sharing and support decision making associated with rapid disease control, prevention, and treatment.

Appendix A

Chapter 4 Glossary

Characterization: Determination of one or more physical, chemical, or biological properties, characteristics, and/or identities of a material and/or biological entity. For example, characterization tests could include determination of agent strain, viability, or transmissibility.

Clinical laboratory: Facility for the biological, microbiological, serological, chemical, immunohematological, biophysical, cytological, pathological, or other examination of clinical samples (see below) to provide information for the diagnosis, prevention, or treatment of any disease or impairment; or of assessing the health of humans, animals, or plants.

Clinical sample: A discrete, unaltered portion taken from a human, animal, or plant for the purpose of examination, study, or analysis to inform diagnosis and treatment.

Detection: Initial determination of the presence or absence of an agent or target in a given matrix; for example, detection of a nucleic acid signature of a known threat in physiological tissues, soil samples, or in an aerosol monitoring system.

Diagnosis: Interpretation of diagnostic (see below) result(s) to inform treatment and control options.

Diagnostic: Reagents, instruments, and systems intended for use in diagnosis of disease or other conditions, including a determination of the state of health, in order to cure, mitigate, treat, or prevent disease or its sequelae.

Non-diagnostic sample: Any sample from the environment, food, plants, animals, or humans analyzed for surveillance or detection purposes but not used in a treatment decision.

Identification: Determination of specific details about an agent present in a given environment, matrix, or clinical sample. For example, identification includes strain classification coupled with expressed plasmids and mapped antibiotic resistance genes.

Laboratory: Facility for the biological, microbiological, serological, chemical, immunohematological, biophysical, cytological, pathological, or other examination of non-diagnostic samples.

Point-of-need (PON): Analysis performed by non-medical personnel in close proximity to sample collection point. For example, field analysis of white powders, use of home pregnancy tests, or surveillance for pathogens and pests at ports of entry.

Point-of-care (POC): Analysis of pathogen presence and/or exposure in the health care environment immediately surrounding a patient. Examples include bed-side tests in a medical unit, ambulance, or mobile transport vehicle, or a physician's office.

Signature: Unique identifying component of a threat (e.g., for biological samples; nucleic acids, proteins, etc.).

Appendix B

BST WG Sub-working group Membership

Aberration Detection

Taha Kass-Hout – CO-CHAIR
Department of Health and Human Services
Food and Drug Administration

Charles Liarakos – CO-CHAIR
National Science Foundation

Russ Bulluck
Department of Agriculture
Animal and Plant Health Inspection Service

Randall Levings
Department of Agriculture
Animal and Plant Health Inspection Service

Geoff Scott
Department of Commerce
National Oceanic and Atmospheric Administration

Julie Pavlin
Department of Defense
Armed Forces Health Surveillance Center

Chris Kiley
Department of Defense
Defense Threat Reduction Agency
Joint Science and Technology Office

Scott Remine
Department of Defense
Joint Program Executive Office for Chemical and
Biological Defense

Seth Foldy
Department of Health and Human Services
Centers for Disease Control and Prevention

Carole Heilman
Department of Health and Human Services
National Institutes of Health

Nancy Jones
Department of Health and Human Services
National Institutes of Health

Patricia Strickler-Dinglasan
Department of Health and Human Services
National Institutes of Health

Samantha Gibbs
Department of the Interior
U.S. Fish and Wildlife Center

Adam Kramer
Department of the Interior
National Park Service

Jonathan Sleeman
Department of the Interior
U.S. Geological Survey
National Wildlife Health Center

Risk Anticipation

Geoff Scott – CO-CHAIR
Department of Commerce
National Oceanic and Atmospheric Administration

Glenn Dowling – CO-CHAIR
Department of Defense
National Center for Medical Intelligence

Adia Bogossian
Department of Agriculture
Animal and Plant Health Inspection Service

Ron Sequeira
Department of Agriculture
Animal and Plant Health Inspection Service

Andrew Wilds
Department of Agriculture
Animal and Plant Health Inspection Service

Jan Carson
Department of Commerce
National Oceanic and Atmospheric Administration

Felix Kogan
Department of Commerce
National Oceanic and Atmospheric Administration

Chris Miller
Department of Commerce
National Oceanic and Atmospheric Administration

Madeline Thomson
Department of Commerce
National Oceanic and Atmospheric Administration

Thiaw Wassila
Department of Commerce
National Oceanic and Atmospheric Administration

Christopher Perdue
Department of Defense
Armed Forces Health Surveillance Center

Roger Breeze
Department of Defense
Defense Threat Reduction Agency
Joint Science and Technology Office

Erica Carroll
Department of Defense
Defense Threat Reduction Agency
Joint Science and Technology Office

Bob Huffman
Department of Defense
Joint Program Executive Office for Chemical and
Biological Defense

Dylan George
Department of Defense
National Center for Medical Intelligence

Nathaniel Hupert
Department of Health and Human Services
Centers for Disease Control and Prevention

Carole Heilman
Department of Health and Human Services
National Institutes of Health

Nancy Jones
Department of Health and Human Services
National Institutes of Health

Patricia Strickler-Dinglasan
Department of Health and Human Services
National Institutes of Health

David Wong
Department of the Interior
National Park Service

Ed Espinoza
Department of the Interior
U.S. Fish and Wildlife Service

John Pearce
Department of the Interior
U.S. Geological Survey

Debra Gulick
Department of State

Gary Resnick
Los Alamos National Laboratory

Helen Cui
Los Alamos National Laboratory

Threat Identification and Characterization

Anne Hultgren – CO-CHAIR
Department of Homeland Security
Science and Technology Directorate

Clay Holloway – CO-CHAIR
Department of Defense
Office of the Assistant Secretary of Defense for
Nuclear, Chemical, and Biological Defense
Programs

Robert von Tersch – CO-CHAIR
Office of the Assistant Secretary of Defense for
Nuclear, Chemical, and Biological Defense
Programs

Randall Levings
Department of Agriculture
Animal and Plant Health Inspection Service

Laurene Levy
Department of Agriculture
Animal and Plant Health Inspection Service

Mike McIntosh
Department of Agriculture
Animal and Plant Health Inspection Service

Scott Cross
Department of Commerce
National Oceanic and Atmospheric Administration

Luther Lindler
Department of Defense
Armed Forces Health Surveillance Center

Eric Van Gieson
Department of Defense
Defense Threat Reduction Agency
Joint Science and Technology Office

Scott Remine
Department of Defense
Joint Program Executive Office for Chemical and
Biological Defense

Richard Kellogg
Department of Health and Human Services
Centers for Disease Control and Prevention

Philip LoBue
Department of Health and Human Services
Centers for Disease Control and Prevention

Sally Hojvat
Department of Health and Human Services
Food and Drug Administration

Patricia Strickler-Dinglasan
Department of Health and Human Services
National Institutes of Health

Carole Heilman
Department of Health and Human Services
National Institutes of Health

Nancy Jones
Department of Health and Human Services
National Institutes of Health

Jessica Appler
Department of Homeland Security
Science and Technology Directorate

Sara Newman
Department of the Interior
National Park Service

Jonathan Sleeman
Department of the Interior
U.S. Geological Survey
National Wildlife Health Center

Sanjiv Shah
Environmental Protection Agency

Information Integration, Analysis, and Sharing

Tom Bates – CO-CHAIR
Lawrence Livermore National Laboratory

Gary Roselle – CO-CHAIR
Department of Veterans Affairs

Shantini Gamage – CO-CHAIR
Department of Veterans Affairs

Lynn Tracey
Department of Agriculture
Animal and Plant Health Inspection Service

Rick Zink
Department of Agriculture
Animal and Plant Health Inspection Service

Scott Cross
Department of Commerce
National Oceanic and Atmospheric Administration

Rohit Chitale
Department of Defense
Armed Forces Health Surveillance Center

John Hannan
Department of Defense
Defense Threat Reduction Agency Joint Science
and Technology Office

Nancy Nurthen
Department of Defense
Defense Threat Reduction Agency Joint Science
and Technology Office

Deena Disraelly
Department of Defense
Joint Program Executive Office for Chemical and
Biological Defense

Karen House
Department of Defense
Joint Program Executive Office for Chemical and
Biological Defense

Jennifer Olson
Department of Health and Human Services
Biomedical Advanced Research and Development
Authority

William Morrill
Department of Health and Human Services
Centers for Disease Control and Prevention

Carole Heilman
Department of Health and Human Services
National Institutes of Health

Nancy Jones
Department of Health and Human Services
National Institutes of Health

Patricia Strickler-Dinglasan
Department of Health and Human Services
National Institutes of Health

Teresa Quitugua
Department of Homeland Security
Office of Health Affairs

Jessica Appler
Department of Homeland Security
Science and Technology Directorate

Charlie Stroup
Department of Veterans Affairs

John Quinn
Department of Veterans Affairs

Ray Arthur
Department of Health and Human Services
Centers for Disease Control and Prevention

Molly Brown
National Aeronautics and Space Administration

Clyde Manning
Office of the Director of National Intelligence

Chris Decker
Office of the Director of National Intelligence